AUSTRALIA'S LAST STEAM RAILWAYS THE SOUTH MAITLAND AND RICHMOND VALE RAILWAYS

JOHN WOODHAMS

First published 2024

Amberley Publishing
The Hill, Stroud,
Gloucestershire, GL5 4EP

www.amberley-books.com

Copyright © John Woodhams, 2024

The right of John Woodhams to be identified as the Author
of this work has been asserted in accordance with the
Copyright, Designs and Patents Act 1988.

All rights reserved. No part of this book may be reprinted
or reproduced or utilised in any form or by any electronic,
mechanical or other means, now known or hereafter invented,
including photocopying and recording, or in any information
storage or retrieval system, without the permission in writing
from the Publishers.

ISBN: 978 1 3981 1021 2 (print)
ISBN: 978 1 3981 1022 9 (ebook)

British Library Cataloguing in Publication Data.
A catalogue record for this book is available from the British Library.

Typeset in 10pt on 13pt Celeste.
Typesetting by Hurix Digital
Printed in the UK.

Contents

	Introduction	4
1	The Discovery of the South Maitland Coalfields and Early Railway Developments	6
2	The South Maitland Railway Era	12
3	J. & A. Brown and the Richmond Vale Railway	49
4	Locomotives and Rolling Stock	63
5	The Heritage Era	83
	Appendix A Locomotive Summary	87
	Appendix B Leading Dimensions of Principal Locomotive Classes	93
	Bibliography	95
	Acknowledgements	96

Introduction

The South Maitland and Richmond Vale Railways together formed an extensive network of standard-gauge lines serving collieries in the New South Wales coalfields. The first section was a railway opened for coal traffic in 1859, between Hexham and a colliery at Minmi owned by John Eales, which subsequently became part of the J. & A. Brown operation and the Richmond Vale Railway (RVR).

The origins of the South Maitland system date back to 1893 and the opening of a line by the East Greta Coal Mining Company to serve its eponymous colliery. The South Maitland Railway (SMR) company was formed in 1918 to amalgamate the East Greta operation with other independent colliery lines that fed into it.

Passenger services were introduced on part of the growing East Greta system from 1902, and such was the volume of traffic that sections of route were later double-tracked and signalled. The New South Wales Government Railways took over passenger services from 1930 until 1961, following which the SMR continued with diesel railcars for a few more years. The RVR also operated passenger trains for miners until 1959.

New shafts were sunk over the years with pits established by a number of companies, which, as might be expected, experienced very mixed fortunes. Some were only in operation for a relatively short period, and all were subject to the variations in the market for coal. This was most noticeably evident in the slump in demand during the 1930s. In the state of New South Wales, it is also perhaps unsurprising that colliery names such as Neath, Aberdare and Abermain could be found.

The railways used a fleet of predominantly British-built steam locomotives, ranging from six-coupled saddle tanks to several 0-6-4T locomotives originally built for use in the Mersey Tunnel, and thirteen tender engines constructed by the British Railway Operating Division (ROD) for use in France during the First World War. However, the last working locomotives were a fleet of Beyer Peacock 2-8-2 tank locomotives, which in later years often worked 1,500-ton trains double-headed. Steam operation ended on the South Maitland system in 1983, leaving the Richmond Vale operation as the last commercial steam working in Australia until that too ceased in 1987, a decade and a half after the end of main-line steam in the state.

The South Maitland Railway company survives as a commercial concern, providing infrastructure support services to the rail industry, including testing and stabling facilities for rolling stock. The cessation of coal production at the Chinese-owned Austar mine at Pelton led to the last section of operational line being mothballed in early 2020. Two years later, the company was acquired by Aurizon Holdings, Australia's largest freight operator. Despite the closure of the Austar mine, and the wider decline of mining in the Hunter Valley area, coal extraction remains an important industry in Australia with 565 million tons produced in 2022.

A section of the Richmond Vale Railway has become a museum line, and fortunately many locomotives from these remarkable operations have survived in preservation.

This book explores the story of the South Maitland and Richmond Vale Railways and their constituents, as well as the heritage era since the end of commercially operated steam, although more detailed histories have been published in Australia. The photographs predominantly illustrate the period from 1970 until 1987, but with a number of archive images dating back to the early days.

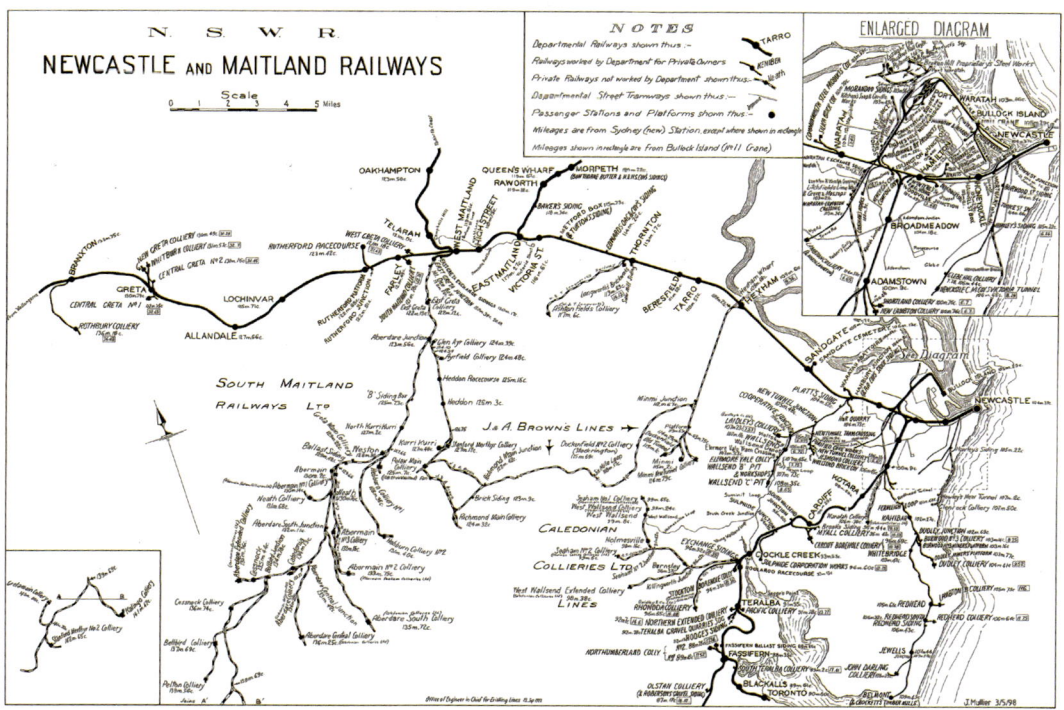

A map showing the South Maitland and J. & A. Brown (Richmond Vale) Railway networks in 1924. (Courtesy Jeff Mullier)

1

The Discovery of the South Maitland Coalfields and Early Railway Developments

The coalfields around the area of South Maitland were the most extensive in New South Wales, originally discovered in 1801 by Lieutenant Colonel Paterson while exploring the Hunter Valley. Although there was no immediate development, several small-scale mining operations were evident by the 1840s. The major breakthrough came in 1886 when Professor Tannatt William Edgworth-David recognised the scale of the deposits, following a find of high-grade coal near the town of Abermain, at Deep Creek, which led to the designation of some 20,000 acres of land for mining.

The Silkstone Coal Mining Company was formed to exploit the Deep Creek find and submitted a proposal to build a railway to connect the colliery area to the nearest main line, the Great Northern Railway, in the Maitland region. Although consent was granted, the line was not built owing to lack of finance, but the rights were acquired by the Clyde Coal, Land and Investment Company. Further mining leases were granted in neighbouring areas to a number of other firms, including Messrs Hebblewhite, Wilkinson & Taylor and Adams & Trenchard.

Another syndicate was formed to purchase an area of land near West Maitland, and this became the East Greta Coal Mining Company in 1891. Although this was a period of financial instability and falling coal prices, the company established sufficient resources to proceed, and reached agreement with the Clyde company to build a section of the original railway proposal which would serve its own needs. Construction work commenced on 20 July 1892, with the line opened to East Greta Colliery the following year. This became known as No. 1 colliery following the opening of another, East Greta No. 2, in 1896. The connection to the main line was about two miles west of Maitland and became known as East Greta Junction.

The East Greta company soon secured further rights from the Stanford Greta Coal Mining Company, and sought to extend the railway to link the new sites. Consent was granted in 1900 with the passing of the Stanford Greta Coal Mining Railway Act, and the new railway was opened in March 1901. Although two undeveloped mines had been acquired, one was

Above: East Greta No. 1 was a 4-4-0T built by Manning Wardle, works No. 39, which had previously been employed by NSWGR Northern Division. (Courtesy University of Newcastle)

Below: An early view of East Greta No. 1 colliery in August 1894. (Courtesy University of Newcastle)

sold to Messrs J. & A. Brown, which then developed it with the name Pelaw Main, and the East Greta company's new facility was renamed Stanford Merthyr. The Stanford Merthyr extension was rather more steeply graded than the original section, with a ruling grade of 1 in 50 approaching J. & A. Brown's Pelaw Main Colliery. A new brickworks at Stanford Merthyr was also linked to the railway by a short spur.

In addition to coal traffic, the railway introduced passenger trains in 1902, using carriages borrowed from the main-line operator. However, the service hardly provided a sensible link for onward main-line travel connections, as only the East Greta company's trains stopped at the junction platform, and Maitland station was over a mile distant – with no proper road link between! From August 1903, the East Greta company's trains worked through to West Maitland station, but, in the meantime, there was much discussion about building a tramway link, even extending as far as Stanford Merthyr. The 'town' of Stanford Merthyr was actually known as Kurri Kurri, an Aboriginal term for 'hurry up', which was served by a separate passenger terminus branching off the colliery line.

The next development was the construction of a line for the Aberdare Collieries of New South Wales Ltd to serve further new collieries in the Cessnock region. Consent was granted in 1901 for a route, some twelve miles in length, with a link to the existing Stanford Merthyr line at a point approximately four miles from East Greta Junction. As the railway was intended for passenger as well as coal traffic, an interchange platform was

Usually known as *Daisy*, No. 1 is seen here on a passenger service in the Hunter Valley in the last years of the nineteenth century. (Courtesy University of Newcastle)

provided at the junction. The section of line from Aberdare Junction to Abermain opened in January 1903, with the onward extension to Cessnock following two years later in February 1904. The line was built using direct labour by the colliery company, and even at this date, largely using bullock teams with ploughs rather than any mechanical excavators for the earthworks required. However, construction on the Abermain portion was assisted by an 0-6-0 tender locomotive dating from 1874, acquired from the Australian Agricultural Company (AAC), while engineers' trains on the Cessnock section were worked by an 1863-built Manning Wardle 0-6-0ST, which had already previously worked for a variety of colliery owners in the Newcastle region.

The AAC was responsible for the development of the Hebburn Colliery, on a tract of land which it obtained from the Aberdare company. A new rail link from the mine to a junction with the Aberdare Railway at Weston (just north-east of Abermain) was in operation by 1903. Three years later the AAC, which already owned 50 per cent of the Aberdare Railway, bought the Aberdare Coal Mining Company's half share, thus giving it full control, and entered into a new agreement with the East Greta company to continue to work the line for a further five years. The level of traffic was by now such that over twenty coal trains operated over the line per day, and it was decided to double the track between East Greta and Weston. The additional capacity was in operation by late 1906, followed by improvements to Aberdare Junction. Remodelling and signalling of the junction at Weston were carried out in 1909. About a mile north of Weston, a passenger halt had been opened in 1904 named North Kurri Kurri.

Avonside 0-6-0ST No. 8 of 1904 heads a train near Heddon Greta, *c.* 1920. (Courtesy University of Newcastle)

Locomotive No. 5, named *New England*, was acquired second-hand in 1903. It was one of three Kitson built 0-6-0s operated by the company. (Courtesy University of Newcastle)

Immediately west of the terminus at Cessnock, the Caledonian Coal Mining Company established the Aberdare Extended Colliery in 1906, having discovered a potentially rich coal seam, 32 feet thick, which remained in production for nearly sixty years. Yet another operator, the Hetton Coal Company, which held mining rights on land south of Cessnock, took the decision in 1909 to build a new line, known as the Hetton Bellbird Railway, from a mine at Bellbird Creek to a junction with the Aberdare line about a mile east of Cessnock.

In 1912 the remaining length of the Aberdare Railway's single-track main line between Neath and Cessnock was doubled. This included the steeply graded Caledonia bank, between Bellbird Junction and Caledonia, where Up trains towards East Greta frequently required banking assistance.

Further developments included a new mine near Aberdare Junction, opened by the Glen Ayr Colliery Company in 1914, and, two years later, the Newcastle-Wallsend Coal Mining Company started work on a mine at Pelton, for which a new railway connection, nearly two miles in length, was required. The new line joined the Bellbird branch near the Hetton Bellbird Colliery sidings.

Passengers waiting on the platform at Cessnock in 1905. (Courtesy University of Newcastle)

However, despite these new developments, the coal-mining industry throughout Australia was struggling with reduced demand following the outbreak of the First World War in 1914. The government banned the export of coal, which, with a slump in domestic demand, resulted in a drop in production from 12.5 million tons in 1913 to under 10 million tons three years later. In 1917 industrial unrest led to a strike by railway workers in the state, which became known as the New South Wales General Strike, as it broadened beyond the transport industry (and also into other states). Employees of the East Greta company joined the strikers, which led to calls for the private railway to be taken over by the state network.

2

The South Maitland Railway Era

The East Greta Coal Mining Company and Hebburn Coal Mining Company agreed to amalgamate their respective railway operations in a new company, South Maitland Railways Ltd (SMR), which was registered and incorporated in 1918. The new entity took over the routes between East Greta and Stanford Merthyr, and Aberdare to Cessnock, as well as the operation of nearly all the remaining colliery feeder lines. At about the same time the Hebburn company opened a new colliery, Hebburn No. 2, about three miles from Weston, and commenced construction of a line from the new mine to a junction with the existing link to its No. 1 colliery near Weston.

Private railways the size of the SMR, with heavy levels of traffic, were comparatively unusual in Australia, and by the mid-1920s, it was handling 46 per cent of the total coal production of the state of New South Wales. It was the most important coal railway in the state, and the significant levels of traffic generated had already, in 1916, led to the government increasing capacity of its main line between Maitland and the port of Newcastle to accommodate it. The only other private railway of comparable size in the state was the 3-foot 6-inch-gauge Silverton Tramway, which over a thirty-six-mile route linked the silver mines of the Broken Hill region to Cockburn in the neighbouring state of South Australia.

The new company adopted the NSWGR signalling system in 1925, with lower quadrant semaphore signal arms as standard, and in a further investment to improve safety a level crossing at Weston was replaced by a new road overbridge at around the same time. In the first eight years of the SMR it continued to take delivery of the, by now, standard class of 2-8-2T locomotives from Beyer Peacock, which are described in more detail in a later chapter.

The years 1925 to 1929 were a period of prosperity for the Maitland region collieries, with over 10,000 men employed producing over 4 million tons of coal each year, but troubled times lay ahead.

By March 1929 industrial relations in the coal industry were at a low ebb when a reduction of 12.5 per cent in miners' wages was announced. This led to a widespread strike, various acts of sabotage with other disruptive action and inevitably a significant loss of traffic for the railway. The company attempted to maintain a service to comply with its legal obligations, but when train crews refused to move stockpiled coal from Stanford Merthyr they were dismissed. A train crewed by management, which took loaded wagons from the colliery, resulted in the closure of the SMR while the miners' dispute became more bitter and prolonged.

The closure also meant the withdrawal of passenger services, and the coach fleet was mostly stabled for the duration in the four-road carriage shed at East Greta Junction. Amid the chaos of the strike, on 1 March 1930 a fire swept through the shed, which destroyed no less than fifteen bogie carriages and over twenty smaller vehicles. Although arson was suspected, it was never proved.

The strike, which became generally known as 'the lockout', had already resulted in the loss of passenger traffic to motor buses, but when compounded by the loss of the carriage fleet, the company decided that it no longer wished to operate its own services. However, the alternative bus services were not proving entirely successful or reliable, and the company negotiated an agreement with NSWGR to operate a passenger service between West Maitland and Cessnock from 13 April 1930, using its own locomotives, rolling stock and crews, leaving the SMR to concentrate on its core business of transporting coal. From February 1940 NSWGR introduced a daily through service between Cessnock and Sydney, which became known locally as the 'Cessnock Express'.

However, by the late 1950s the passenger operation was incurring substantial losses for the SMR, and the company sought to terminate its contract with the state railway. The government-run service ended in 1961 when the SMR introduced three new diesel railcars, with the hope that a lower cost service would be sustainable. The 'Cessnock Express' continued for the time being, although by now only running through to Broadmeadow, on the outskirts of Newcastle, rather than Sydney. However, the decline in traffic continued, leading to the withdrawal of Saturday afternoon and all Sunday services in May 1965. This measure did not lead to a turnaround in fortunes as hoped for, and all passenger workings ended on 24 January 1967.

Following the disruption caused by the 1929–30 strike, the coal industry suffered several years of depression, which inevitably led to the closure of not only many collieries, but the loss of the market between Aberdare Junction, Kurri Kurri and Stanford Merthyr. Several smaller and older locomotives which thus became redundant were then sold. This line also suffered from subsidence, caused by an underground fire, from the Ayrfield Colliery about a mile south of Aberdare Junction, and it was proposed to build a new link between Weston and Pelaw Main Colliery. Although not a new idea, the subsidence created a certain urgency for a viable route to Kurri Kurri, and work commenced on the new link in May 1936 with completion just six months later. In theory the new link provided J. & A. Brown with the potential to abstract coal traffic from Abermain and Stanford via its own Richmond Vale Railway to Hexham (see next chapter), a proposal that encountered trade-union opposition. However, gas coal from J. & A. Brown mines was taken by SMR locomotives to Pelaw Main, and thence worked by the RVR to Hexham, but all Newcastle-bound coal was worked by the SMR to East Greta. The link provided a useful interchange between the two concerns until rail traffic at Abermain No. 2 colliery stopped at the end of 1963, when it fell into disuse.

The East Greta Coal Mining Company was taken over by J. & A. Brown and Abermain Seaham Collieries (JABAS) in April 1931, which thus gained a 50 per cent stake in the SMR. The legal entity of the SMR became South Maitland Railways, Proprietary, Ltd in the following decade to comply with changes in company regulation.

The outbreak of the Second World War led to an increase in demand, and before very long forty loaded coal trains were being worked each day between Cessnock and East Greta, with the South Maitland mines now producing 60 per cent of northern NSW's total output. A number of locomotives which had been laid aside awaiting repair during the years of the Depression were overhauled as quickly as possible in order to provide sufficient motive power for the upsurge in traffic.

Coal was conveyed in unfitted timber wagons, and, later, NSWGR air-braked hopper wagons, which could be worked in train loads of up to 1,000 tons. It was not unusual for two Down (empty) trains to be coupled together to provide sufficient paths within the constraints of the signalling layout, and, conversely, it was sometimes necessary to split a loaded train into two parts for the ascent of Caledonia Bank.

From time to time the railway suffered from flooding of the Hunter River, and in June 1949 the SMR was brought to a halt with most of its locomotives stranded in floodwater at the East Greta depot. Yet again, in 1955, the railway was largely out of action from 25 February until 8 April, with at one time over 5 feet of water in the engine shed. With much of the locomotive fleet stranded, available engines were serviced at the RVR Pelaw Main shed, and others were hired in from the state railway system. Earlier floods in 1930 had caused severe damage to underground workings in the region, as well as undermining the railway infrastructure. In addition to flooding, there was further disruption in 1949 caused by another strike in the mining industry, which lasted seven weeks, with the government bringing in troops to work the open-cast pits in an attempt to break the dispute. Indeed, the miners returned to work, defeated, two weeks later.

In 1960 ownership of the SMR changed once again, as JABAS merged with Caledonian Collieries to form Coal and Allied Industries, and its 50 per cent share passed to the new entity. Coal and Allied's acquisition of Hebburn Ltd in 1967 then gave it full control of the SMR.

Rationalisation within the coal industry continued throughout the 1960s. The pithead site at Abermain No. 1 colliery was closed as its underground workings amalgamated with No. 3 pit, but that too was closed in 1960, leaving only Abermain No. 2 colliery until that was shut completely in 1964. Aberdare Colliery at Caledonia and the Aberdare Extended Colliery at Cessnock also closed, but, despite this retrenchment, in 1967 around thirty loaded coal trains, as well as the equivalent number of empty return journeys, were worked over the SMR on a daily basis.

The end of steam operation on the state network came in early 1973, but the SMR steam fleet remained at work for a further decade. Indeed, as late as 1981 the SMR opened a new locomotive shed at East Greta. On 29 May 1982 locomotives 17 and 31 working a double-headed train of government-owned bogie hoppers became derailed near Kurri Kurri with both locomotives ending up on their sides on the embankment. The state railway's breakdown crane was called in to lift the two engines and clear the wreckage, and both engines were repaired and returned to service. The railway was still busily engaged transporting the coal output from three working collieries, but the end was not far off, and steam operation ended on the old SMR lines just over a year later on 10 June 1983, with state railway diesel locomotives taking over the traffic the following day. However, East Greta shed would still be required to provide steam locomotives – some now working into their seventh decade – for the Hexham-based Richmond Vale line for a further four years.

No. 19 at East Greta shed, 13 April 1968. (Courtesy Graeme Kaufman)

Above: Nos 19 and 25 await their turns at East Greta shed, 13 April 1968. (Courtesy Graeme Kaufman)

Below: Nos 17 and 22 simmer quietly at East Greta during an autumn evening in March 1978. (Courtesy Dennis Rittson)

Above: No. 28 by the East Greta coaling stage, *c.* 1974. (Courtesy Dennis Rittson)

Right: A feature of East Greta depot was the Sunday shunt, which involved coaling and positioning locomotives ready for the following week. In December 1978 No. 20 propels Nos 19, 31 and 24 back to the shed ready to be lit up. (Courtesy Brian Ayling)

Above: No. 20 carefully positions No. 24 under the hopper, to be coaled ready for its next duty. (Courtesy Brian Ayling)

Below: A general view of the shed yard at East Greta, 1976. (Courtesy Ian Lynas)

Above: The interior of the new shed opened in 1981. (Courtesy Warren Dibb)

Below: Work to a buffer beam bracket at Mount Dee works, East Greta, October 1979. (Courtesy Dennis Rittson)

Above: The tank of an unidentified 10 class receives works attention. (Courtesy Dennis Rittson)

Below: Work is underway on a front tubeplate in the boiler shop. (Courtesy Dennis Rittson)

Above: The SMR works was capable of heavy boiler repairs including fabricating new fireboxes. (Courtesy Dennis Rittson)

Below: No. 20 awaits the fitting of replacement cylinders in March 1980. (Courtesy Dennis Rittson)

Above: 4-6-4Ts Nos 15 and 29 derelict by Mount Dee signal box, East Greta, 13 April 1968. (Courtesy Graeme Kaufman)

Below: Mount Dee signal box, April 1968. At its peak the railway had thirteen signal boxes in use, but in later years only two were required. (Courtesy Graeme Kaufman)

Above: An early morning scene as class 10 2-8-2T No. 27 is prepared for the day. (Courtesy Dennis Rittson)

Below: No. 22 with an early morning departure from East Greta sidings with empties, August 1979. (Courtesy Dennis Rittson)

Above: Nos 10 and 22 climb past Swamp Creek ponds with empties for Pelton Colliery, October 1979. (Courtesy Dennis Rittson)

Below: No. 10 with empties at Gilleston Heights, October 1977. (Courtesy Dennis Rittson)

Above: An early morning double-header at Swamp Creek, October 1979. (Courtesy Dennis Rittson)

Below: No. 26 catches the early morning sun at Swamp Creek, October 1979. (Courtesy Dennis Rittson)

Above: No. 20 and state railway No. 4831 in the East Greta Junction exchange sidings, September 1982. (Courtesy Brian Ayling)

Below: NSWGR D53 class 2-8-0 No. 5262 heads away from Maitland on the main line to Port Waratah with a loaded coal train from the SMR in April 1972. (Courtesy Dennis Rittson)

Above: No. 10 with an unfitted train at Gilleston Heights, 11 May 1973. (Courtesy Dennis Rittson)

Below: An unfitted Up train at Gilleston Heights, 11 May 1973. (Courtesy Dennis Rittson)

Above: Nos 10 and 31 work through Weston in the last week of SMR steam operation, 2 June 1983. (Courtesy Dennis Rittson)

Below: No. 28 passes through Weston station bound for Bellbird Colliery in 1974. (Courtesy Dennis Rittson)

Above: Nos 24 and 18 storm through Weston with a train from Pelton in the last days of SMR steam, June 1983. (Courtesy Dennis Rittson)

Below: A general view of Weston station in 1965. (Courtesy Lindsay Bridge)

Above: No. 17 takes a water stop at Weston with a load from Neath to East Greta, 8 May 1979. (Courtesy Brian Ayling)

Below: On 29 May 1982 Nos 17 and 31 were derailed by a herd of cattle on the line near Kurri Kurri. (Courtesy University of Newcastle Phillip Lockett collection)

Above and below: No. 31, the pilot engine, is lifted back onto the track the following day by a crane from the state railway. It was subsequently repaired using a number of parts from sister No. 19. (Courtesy University of Newcastle Phillip Lockett collection)

Above: Nos 30 and 31 approach Neath Junction signal box, with signalman and fireman ready for the single-line staff exchange, 7 June 1983. (Courtesy Dennis Rittson)

Below: No. 26 passes the site of the former junction linking Pelaw Main in 1979. (Courtesy Brian Ayling)

Above: Token exchange at the Cessnock Road level crossing on the Neath Colliery branch, January 1979. (Courtesy Dennis Rittson)

Below: No. 18 and van have been attached to a Pelton-bound train hauled by Nos 10 and 27, 7 September 1978. (Courtesy Brian Ayling)

Above: Nos 17 and 30 cross Swamp Creek at Abermain with a ballast train working in 1980. (Courtesy Dennis Rittson)

Below: Nos 18 and 22 climbing Neath Bank with unfitted hoppers for Aberdare Colliery, September 1977. (Courtesy Dennis Rittson)

Above: Nos 22 and 18 approach Caledonia yard bound for Aberdare washery, September 1977. (Courtesy Dennis Rittson)

Below: Both engines are detached to take water at the Caledonia water crane. (Courtesy Dennis Rittson)

Above: Nos 30 and 22 at Neath Bank heading for Pelton in August 1979. (Courtesy Dennis Rittson

Below: No. 30 shunts at Aberdare washery before coupling to its loaded train ready to return to East Greta, August 1979. (Courtesy Dennis Rittson)

Above: With a train of 'Black Ducks' for Bellbird Colliery, No. 31 approaches the summit of the climb to Abermain, May 1974. (Courtesy Dennis Rittson)

Below: No. 30 with coal hopper modified for ash ballasting with plough and automatic couplings fitted, 15 May 1979. (Courtesy Brian Ayling)

Left: The ballast gang at work in Denman Hill cutting, 15 May 1979. (Courtesy Brian Ayling)

Below: No. 30 has reattached the ballast hopper to its loaded coal train, and heads across Swamp Creek Bridge, Abermain, en route to East Greta. (Courtesy Brian Ayling)

Above: Nos 22 and 30 charge the 1 in 70 gradient of Caledonia Bank in August 1979. (Courtesy Dennis Rittson)

Below: On a cold winter morning Nos 10 and 20 storm Caledonia Bank with an Up coal train. (Courtesy Dennis Rittson)

Above: No. 24 with empties waits for the road near Caledonia station as a double-headed Up train passes, July 1978. (Courtesy Dennis Rittson)

Left: SMR No. 15, a Beyer Peacock product of 1912, taking water at Cessnock, 23 February 1961. (Courtesy the late Weston Langford)

Above: SMR 4-6-4T No. 15 meets NSWGR No. 3125 at Cessnock in early 1961, the final year of steam-hauled passenger services. (Courtesy the late Weston Langford)

Below: NSWGR No. 3125 runs round its train at Cessnock, 23 February 1961. (Courtesy the late Weston Langford)

Above: No. 24 with a load from the coal washery waits at Aberdare for a Pelton-bound double-header to pass, July 1978. (Courtesy Dennis Rittson)

Below: A coal train from Pelton headed by Nos 26 and 24 passes the remains of Bellbird Colliery, 7 September 1978. (Courtesy Brian Ayling)

Above: Nos 22 and 18 approach Aberdare coal washery, September 1977. (Courtesy Dennis Rittson)

Below: The track gang appears unconcerned by the approach of Nos 20 and 26, near Aberdare washery, July 1978. (Courtesy Dennis Rittson)

Above: A night departure from Aberdare washery for No. 27 in March 1978. (Courtesy Dennis Rittson)

Below: In July 1978 No. 27 was derailed during shunting operations at Aberdare washery. (Courtesy Dennis Rittson)

Above: No. 26 pilots No. 17 as they start a train from Pelton yard in July 1978. (Courtesy Dennis Rittson)

Below: Nos 27 and 10 loading their train at Pelton by floodlight in May 1978. (Courtesy Dennis Rittson)

Above: A train loading at Pelton Colliery on the afternoon of 8 May 1979. (Courtesy Brian Ayling)

Below: In the last week of steam operation at the SMR No. 17 pilots an unidentified train engine, 5 June 1983. (Courtesy Ian Lynas)

Above: Two days before the end Nos 17 and 22 head towards East Greta with a loaded train. (Courtesy Ian Lynas)

Below: No. 31, adorned with a wreath, heads an Up train to East Greta on the last day, 10 June 1983. (Courtesy Ian Lynas)

Above: The last steam working to return to East Greta, No. 22 with van, 10 June 1983. (Courtesy Ian Lynas)

Below: No. 22 returns to the shed yard, its working days over. (Courtesy Ian Lynas)

3

J. & A. Brown and the Richmond Vale Railway

Brothers James and Alexander Brown began small-scale mining operations at East Maitland in 1843, moving to a new area near Newcastle in 1852. The following year they bought more land at Minmi, where John Eales had already established mines and built a railway to Hexham. They sunk their first shaft at Minmi in 1857, but could not reach an agreement with Eales to transport their own coal production, and two years later bought out his operation for £41,000. A third pit was established at Minmi in 1861, along with new workshops to service the mines and the railway.

The Minmi mines only had a short life before closure in 1869, by which time the Brown family had established new mining interests at New Lambton. However, in 1874 James Brown and his son John opened a new mine, Duckenfield Colliery, near the site of the earlier third mine, known as C pit, at Minmi. A second mine, known as Brown's, or sometimes Back Creek Colliery, commenced production in early 1877.

The Richmond Vale Colliery near Kurri Kurri was initially sunk by a mining syndicate, but it was acquired by the Brown family in 1897 and further developed. In 1900 the company sought parliamentary approval to build a railway to link the Richmond Vale site with its existing Minmi to Hexham line. In October 1900 the Richmond Vale Coal Mine Railway Act was passed, but in the same month the Browns bought the Stanford Greta Colliery, which they decided could be worked more economically than the Richmond Vale mine.

The Stanford Greta mine was renamed Pelaw Main, and initially the coal produced was transported by the East Greta railway. However, a short branch was built from the new Richmond Vale line to serve Pelaw Main, and that traffic was lost to the East Greta concern. When John Eales had originally built the line between Minmi and Hexham he had established a wharf and loading facilities on the Hunter River at Hexham, so coal traffic from Pelaw Main could thus be taken directly to a port using only the firm's metals. Work started on the new line in 1904, and it was completed to Pelaw Main in June of the following year, and to Richmond Vale Colliery two months later. With the railway built there was renewed interest in developing the Richmond Vale site. With substantial investment, and renamed Richmond Main, by 1914 a new No. 2 shaft was ready for production. A brickworks was established for the construction of the pithead buildings, which included a substantial electricity power station. Other facilities included a locomotive shed, which was ready in 1916. Despite the hopes for the colliery, which the firm regarded as a state-of-the-art showpiece, production in the early years was beset by industrial problems, but in 1928, the year of peak production, 507,000 tons of coal were raised, and 1,200 men were employed.

As the old Duckenfield and Brown's collieries became exhausted a new pit was sunk at Stockrington in 1912, named Duckenfield No. 2, although it was not brought into use immediately. Although further work was carried out during the 1920s, by which time the older pits had closed, it did not enter full production until as late as 1936, long after the industrial unrest and lockout of 1929–30. It was by now renamed Stockrington Colliery. Further mines opened later, with Stockrington No. 2 colliery, No. 3 Tunnel starting production as late as 1954. An open-cast mine had opened on the site of the old Minmi pits in 1949, but that closed in 1954.

In 1922 a new line was built between Pelaw Main Colliery and Richmond Main Junction, which was used for the operation of miners' passenger trains, and also increased capacity on the system. The miners' service continued to run until withdrawn in 1959. Locomotive operations were centred at Pelaw Main from 1925.

In January 1931 a new company was formed, J. & A. Brown and Abermain Seaham Collieries Ltd (JABAS). Although long-serving John Brown had died the previous year, the family was still actively involved in the running of the new company with the appointment of Stephen Brown as a director. As described in the previous chapter, this gave the company a 50 per cent stake in the SMR.

In 1953 the company built a coal-grading plant and washery at Hexham to cater for an increase in demand for smaller grades of coal. Located adjacent to the exchange sidings, the new facility had rail connections to both the RVR and NSWGR main line. However, within a year or two of the opening of the new Hexham plant there was an overall slump in demand, leading to the closure of a number of collieries in the Stockrington area. Stanford Main No. 1 also closed in 1957 followed by Pelaw Main in 1961, although the locomotive shed at the latter remained in use. Richmond Main, which had once been the showpiece for the Brown empire, and at one time the most productive mine in the state, closed in July 1967, although the power station at the site remained in use until 1976. This left Stockrington No. 2 as the main source of coal, and the line beyond it was closed. Much of the route to Hexham had been doubled in more buoyant times, but the remaining line was now reduced to single-line working with the redundant track used for storage of large numbers of surplus wagons. The railway, which had been part of Coal and Allied Industries since 1960, was regularly working seven or eight train loads daily from Stockrington No. 2.

Track beyond the closed Stockrington No. 1 colliery to Richmond Main was lifted in 1973, along with the Minmi branch, which had been out of use for many years. The line beyond Hexham exchange sidings to the wharf, including the level crossing with the government main line, which had last been used in 1967, was also dismantled. Reusable materials from the demolition were reclaimed for future use.

A high proportion of trains from the Maitland coal industry consisting of older unfitted wagons were still being worked over the main-line network, in particular to Port Waratah, but in 1972 the newly formed Public Transport Commission (PTC), which had assumed responsibility for the state system, announced that it intended to ban the use of non-air-braked stock on its tracks. However, the PTC agreed to accept the continued working of non-air-braked stock between East Greta, the SMR's junction with the main line, and Hexham, where the RVR had always had its own exchange sidings and junction. Coal and Allied therefore decided to invest in a new Unit Train loading facility at Hexham, the location of the existing preparation plant and washery, which handled all coal leaving its railway. A new balloon loop was built, using trackwork reclaimed from the recent demolition, to serve the new loading plant, which had a capacity of 1,200 tons. The last unfitted coal train was despatched to Port Waratah on 15 August 1973, two days before the opening of the new loader. However, the PTC still wanted to eliminate the non-fitted workings between East Greta and Hexham, and announced its intention to cascade older bogie hopper wagons, classified BCH, to the SMR. A start was made on fitting buckeye couplings to locomotives in order that they could work the new stock, which was scheduled to be introduced at Easter 1977. However, insufficient wagons were made available, and as many older types had already been withdrawn, there was a severe shortage of serviceable stock for some months, which led to additional weekend workings. The last unfitted train between East Greta and Hexham ran in February 1978, although older wagons still remained in use at the RVR.

Although steam operation ended at the SMR in 1983, the 10 class 2-8-2Ts could still be found at work between Stockrington and Hexham until 1987 – the very last commercial use of steam locomotives in Australia. At the start of the decade steam locomotives could also be found in industrial service at the Blue Circle Southern Cement Company's works at Portland. They were used to shunt the factory and work trains over a short but steeply graded branch line from the main-line exchange sidings. Three British-built engines, two standard 0-6-0Ts by Andrew Barclay, and a rather more unusual ex-NSWGR 2-6-2ST built by Dübs in 1891 were still at the works when diesels took over in August 1982. The works and railway closed completely in 1991.

Coal and Allied then decided that in the prevailing economic environment, road transport would be a more cost-effective operation than retaining the railway, and the only way to keep the Stockrington mine viable. On 28 August 1987 the company announced that the railway would close on 25 September – just four weeks later. However, the closure did not happen quite as envisaged by management, as on 18 September the operation of loaded coal trains ceased abruptly following a dispute with the footplate crews. For a few more days empty wagons were worked to Hexham for storage, and although some coal arrived at the washery on 21 September, the plant itself was blockaded by locomotive No. 30 and a rake of wagons to prevent the loading of lorries which had already been introduced. On the following day No. 25 undertook two trips light engine to Stockrington to retrieve more empties for storage at Hexham. Then in a further bizarre action by staff, some of whom had been summarily dismissed, on 24 September the engine was driven to Doghole, just short of the Stockrington terminus, and thence to a bridge at Lenaghans Drive, where a protest camp was established, ironically within sight of the lorries now transporting coal from the mine. A nearby farm provided water, and a supportive lorry driver left a quantity of coal, allowing No. 25 to be kept in steam. The anticipated wider trade-union support for the railway staff did not materialise, and the thoroughly disheartened RVR employees capitulated on 15 October. And with the return of No. 25 to Hexham, recorded by a TV film crew, the era of working steam locomotives on the continent came to an end.

Had the decision to opt for road transport not been taken in 1987, the railway could only have lasted for a few more months, as the washery and processing plant was closed in May 1988, followed a few weeks later by the closure of Stockrington No. 2 colliery. By that time rails had already been lifted in the colliery area, and most of the 700 wagons remaining on the railway had been scrapped. By the end of the following year both the colliery and preparation plant had been demolished, and the engineering workshops at Hexham, which built and maintained mining equipment, were closed. This spelt the end of operations in this area for Coal and Allied, which was subsequently taken over by H. & M. Holdings, Proprietary, Ltd in 1989.

Kitson-built 2-8-2T No. 10 *Richmond Main* propels a train of wagons across the NSWGR main line at Hexham towards the wharf, 28 November 1967. (Courtesy the late Weston Langford)

Above: NSW state railway 2-8-0 No. 5164 eases an Up coal train from the Hexham exchange sidings onto the main line, 20 January 1962. (Courtesy the late Weston Langford)

Below: Kitson No. 10 *Richmond Main* at Hexham, 13 April 1968. (Courtesy Graeme Kaufman)

Above: Kitson tank No. 9 *Pelaw Main* shunts ROD No. 20 at Hexham, 5 May 1976. (Courtesy Dennis Rittson)

Below: ROD No. 17 shunts at Hexham, 24 March 1965. (Courtesy the late Weston Langford)

The driver of ROD No. 17 oils round while sister locomotive No. 19 shunts the yard, March 1965. (Courtesy the late Weston Langford)

No. 25 is serviced at Hexham, 22 November 1985. Although fitted with buckeye coupling, it still retains buffers and three-link coupling as well. (Courtesy the late Weston Langford)

SMR No. 23 shunting at Hexham, May 1978. (Courtesy Dennis Rittson)

Above and below: SMR No. 20 propels a load towards the unloader at Hexham, May 1978. (Courtesy Dennis Rittson)

Above: No. 23 crosses the Hexham swamps with a load from Stockrington, August 1979. (Courtesy Dennis Rittson)

Below: No. 15 with an Up coal train at Stockrington, 9 January 1962. (Courtesy the late Weston Langford)

Above: No. 24 at Doghole, near Stockrington, 3 December 1971. (Courtesy University of Newcastle Philip Lockett collection)

Below: ROD No. 23 with an Up coal train at Stockrington, 24 March 1965. (Courtesy the late Weston Langford)

Above: No. 17 shunts at Stockrington, March 1965. (Courtesy the late Weston Langford)

Below: No. 25 shunts empties at Stockrington No. 2 colliery, August 1978. The wagons would then be hauled through the loading plant by cable. (Courtesy Dennis Rittson)

Above: Shunting at Stockrington, July 1978. (Courtesy Dennis Rittson)

Below: No. 15 departs from Pelaw Main, 19 January 1962. (Courtesy the late Weston Langford)

Above: A general view of Pelaw Main shed, 1959. (Courtesy Lindsay Bridge)

Right: Steam locomotives were still at work at the Portland cement works until August 1982. A remarkable survivor still earning its keep in January 1981 was former NSWGR 2-6-2ST No. 2605. Built in Glasgow by Dübs in 1892 and eventually sold for industrial use in 1966, the veteran is seen shunting the works yard. (Courtesy Frank Hinde)

Portland cement No. 3 was built in 1911 by Andrew Barclay, works No. 1234, seen here in 1978. (Courtesy Brian Ayling)

4
Locomotives and Rolling Stock

East Greta Coal Mining Company and South Maitland Railway

Nearly all the locomotives of the East Greta/South Maitland and J. & A. Brown/Richmond Vale systems were British-built, both to main-line and standard and industrial designs, although a few locally built examples found their way on to the colliery railways.

The first locomotive used on the East Greta railway was an outside cylindered 4-4-0T which had been built by Manning Wardle in 1862. It was used on the construction of the line and had formerly been used on the NSWGR. Appropriately allocated the number 1, it was withdrawn in 1911.

East Greta No. 1 was a 4-4-0T built by Manning Wardle, works No. 39, which had previously been employed by NSWGR Northern Division. (Courtesy University of Newcastle)

An Avonside 0-4-0ST built in 1900 was supplied to the East Greta company as its No. 2. It is seen here at Mount Dee sidings, East Greta, shortly before sale to J. & A. Brown in 1934. (Courtesy University of Newcastle)

Locomotive No. 2 was a four-coupled saddle tank built by Avonside in 1900, and in its early years was used predominantly for working on the extended line to Stanford Merthyr mine. It was sold to J. & A. Brown in 1934, and still survives at Dorrigo Steam Railway and Museum.

No. 3 was a larger six-coupled Avonside product of 1902, while Nos 4 and 5 were inside-cylindered, six-coupled tender locomotives intended for heavier coal trains. Built by Kitson in 1877/9, they were acquired second-hand from contractors engaged on NSWGR construction contracts in 1902 and 1903 respectively. At some point in their careers, they had been given the names *Gang Forward* and *New England*. They were joined by a third, similar engine in 1904, which was numbered 7, bearing the name *Murrumbidgee*. All three were withdrawn by 1922, but they languished at East Greta Junction until cut up in 1927.

In 1903 the company had ordered another new locomotive from Avonside, but this time a larger eight-coupled saddle tank, No. 6, which entered service in January the following year. Two more saddle-tank locomotives followed from the Bristol works of the Avonside Engine Company in 1904, one six-coupled, No. 8, and another eight-coupled example, No. 9. Nos 3 and 8 were often allocated passenger duties, while Nos 6 and 8, by far the most powerful locomotives in the fleet, worked heavy trains from Aberdare Colliery. No. 8 was sold to Gunnedah Colliery in 1928, while during the Depression of the 1930s, No. 9 was sold to the Mount Kembla Coal Company.

Locomotive No. 10 was actually a locally built machine, an 0-6-0 built by Vale and Lucy at Sydney in 1873. It had been NSWGR No. 21, and although withdrawn as early as 1892, it was not acquired by the East Greta concern until 1906, and apparently only worked there for a year or two until it was condemned.

Above: No. 4, built in 1877, was one of a trio of 0-6-0s built by Kitson acquired second-hand in 1902–04. (Courtesy Picture Maitland Collection)

Below: Built in 1877 by Kitson of Leeds, No. 4 was purchased by the East Greta company in 1902. (Courtesy University of Newcastle)

Above: No. 7 *Murrumbidgee* was an 1879 product of Kitson, originally supplied to Amos & Co. (Courtesy University of Newcastle)

Below: In 1903 the East Greta railway ordered an eight-coupled saddle tank, which became No. 6, photographed here with coupling rods to rear axle removed. (Courtesy University of Newcastle)

East Greta No. 9 was an Avonside 0-8-0ST, ordered in 1904. (Courtesy University of Newcastle)

Two 4-4-2T locomotives were acquired from NSWGR in 1907 for working passenger trains. Heavier bogie carriages had been acquired which necessitated the search for more suitable motive power. Built as 4-4-0 tank locomotives by Beyer Peacock in 1877, they were originally very similar to the Metropolitan A class used in the UK. After twenty years' service they were rebuilt with a trailing axle as the NSWGR's CC class. Nos 83 and 84 became East Greta's Nos 12 and 11 respectively, and they both continued at work until withdrawal in 1935.

In 1908/9 the company took delivery of two 0-8-2Ts from Avonside, the first from this builder with side rather than saddle tanks. Numbered 13 and 14, they were acquired specifically for working heavy coal trains on the steeply graded Caledonia Bank between Cessnock and Caledonia. Both engines were redundant by the mid-1930s, and although No. 14 found a new owner, Hetton Bellbird Coal Company in 1936, No. 13 remained out of use until 1945 when it was eventually sold for further use at Newcastle Steel Works. Here, it was extensively rebuilt as an 0-6-0T, renumbered 27 in the steelworks fleet. No. 14 is preserved at Dorrigo. Another Avonside 0-8-2T followed in 1911, numbered out of sequence as No. 1 to replace the recently withdrawn Manning Wardle 4-4-0T. The replacement No. 1 was sold in 1937 to Bulli Colliery. The three Avonside 0-8-2Ts were fitted with large 20 x 24 inch outside cylinders and 3-foot 10-inch diameter coupled wheels. A fairly large grate of 20.7 square feet was provided, although boiler pressure was relatively low at just 140 lb.

One of a pair of 0-8-2Ts ordered from Avonside, No. 14 is seen when still fairly new in 1908. (Courtesy University of Newcastle)

Latterly the SMR became synonymous with one type of locomotive in particular, the class 10 2-8-2Ts, of which fourteen were built between 1911 and 1925. By the end of the first decade of the twentieth century, the East Greta locomotive fleet was barely able to cope with increasing train weights and the demands placed upon it. Therefore, it was decided that the company required new locomotives with similar capabilities and power of the NSWGR engines which regularly worked the coal traffic onward from the exchange sidings. Based on two existing NSWGR classes of tender locomotive, the C32 or P6 4-6-0 and D50 (T524) 2-8-0, the company developed a new tank locomotive design, with many common dimensions and components, but a 2-8-2T configuration. Hitherto, all new locomotive orders had been placed with Avonside Engine Company of Bristol, but this time Beyer Peacock were awarded the contract. The locomotives were shipped from Manchester to Australia in a partially dismantled form, and assembled at East Greta works. They were allocated numbers 10, 17–20, 22–28, 30 and 31. In order to cope with tight curves, both intermediate and driving axle wheels were flangeless. The boilers were pressed to 180 lb and fitted with Belpaire fireboxes.

The first engine delivered, No. 10, was initially used for assisting trains climbing Caledonia Bank, but its capabilities were soon put to full use working unfitted trains of up to 610 tons, with the maximum increased to 698 tons for fitted wagons. The latter was increased to 750 tons between Caledonia and East Greta.

Several locomotives requiring major boiler work were laid aside in the 1950s following a downturn in coal demand, but as the company's prospects improved in the following decade a programme of overhauls and heavy repairs was put in place. This coincided with the takeover of the SMR by Coal and Allied Industries, which already owned the former J. & A. Brown system, and two 10 class engines were transferred to Hexham in 1967, leading to gradual integration of operation of the two networks.

Known by enginemen as 'Bobtail Ts', the 10 class maintained services until the end of steam operation on the SMR in June 1983, when No. 31 took the dubious honour of hauling the last revenue-earning steam-hauled train. Just one year earlier, the engine had been derailed, piloting a double-headed train with No. 17, both locomotives ending up on their sides. The locomotives were normally finished with black paintwork lined in red, but in the final years a few engines were turned out in a green or blue livery.

However, it was not quite the end as steam operation continued at the former J. & A. Brown line between Stockrington and Hexham until that too came to an end in September 1987. Four locomotives were regularly required, with the remainder placed in store, for what was now the last bastion of commercial working steam in Australia. Locomotives were regularly returned to East Greta for maintenance, being hauled to and from Hexham along the main line by state railway diesels. The operation attracted much attention in its last years, but as described earlier, the closure, when it came, did not take place without incident!

Following the order for the first 10 class locomotives, the East Greta Company awarded Beyer Peacock a further contract for two 4-6-4T engines for passenger traffic. Numbered 15 and 16 (Works Nos 5603 and 5638 respectively) were very similar to the NSWGR's S636 (later C30) class, the first of which had also been built by the Gorton firm in 1903/4. The two new engines, delivered in 1912, worked the West Maitland–Cessnock services, with a

A Beyer Peacock works photograph of a class 10 2-8-2T, of which a total of fourteen were ordered between 1911 and 1926. (Courtesy ETH Archive)

The 2-8-2Ts were delivered in knocked down form, and an as yet un-numbered example is seen nearing completion at East Greta. (Courtesy Picture Maitland Collection)

third engine, No. 29, delivered in 1923, until the passenger operation passed to NSWGR in 1930. Thereafter they were used for coal traffic from Abermain and Hebburn. A fourth engine of the type, but this time ex-NSWGR, No. 3013, was acquired on withdrawal as late as 1967, and although this example has survived into preservation, Nos 15, 16 and 29 were scrapped in 1973. No. 16 had actually been withdrawn as early as 1946.

After the integration of the East Greta and Hebburn mining companies in 1918, the respective railway operations were combined in the newly formed South Maitland Railways, Proprietary, Ltd (SMR). The new company decided that it needed another engine for passenger work, and an M40 4-4-2T, No. 51, was acquired from NSWGR in 1919. This was also a Beyer Peacock product, built as part of an order for fifteen locomotives in 1891. It became SMR No. 21, but was laid aside during the 1930s, and eventually sold to the New South Wales Public Works Department in 1941 for shunting at Port Kembla.

In 1903 the Hebburn company acquired two second-hand locomotives for use at the Weston sidings, No. 2, a six-coupled saddle tank built by Beyer Peacock in 1884, and No. 62XX, another 0-6-0 but a product of Robert Stephenson in 1874. No. 2 came from the AAC's Newcastle operation, while No. 62XX was originally an NSWGR engine, but bought from the Newcastle Coal Mining Company. No. 62XX was withdrawn in 1927, but No. 3 served for another decade and was not scrapped until 1947.

Above: No. 15, Beyer Peacock works No. 5603 of 1912, was identical to the NSWGR S636 (later C30) class, intended for passenger workings. (Courtesy Lindsay Bridge)

Below: No. 15, shipped to Australia in knocked down form, during assembly at East Greta in 1912. (Courtesy Picture Maitland Collection)

At East Greta Junction, *c.* 1920, is No. 21, an ex-NSWGR 4-4-2T of 1891 that was acquired in 1919. (Courtesy University of Newcastle)

In 1918 another Beyer Peacock 0-6-0ST was transferred from the Newcastle collieries to help with additional traffic generated by Hebburn No. 2 colliery. It had been built in 1903, and became No. 3 in the Hebburn fleet, where it only remained for two years, being sold to Hoskins Ironworks, Lithgow, in 1920. The sale of this locomotive resulted in the need to hire in additional motive power, and very shortly afterwards an E10 (later Z20) class 2-6-4T arrived from NSWGR, followed by the rather smaller SMR 0-4-0 No. 2.

For a brief period another former NSWGR locomotive was used, Beyer Peacock built 2-4-0T No. 356X. A member of the F351 (later X10) class, the type was originally ordered for working Sydney suburban services, but following a serious derailment in the early years of the twentieth century it was relegated to shunting and yard duties.

In 1934 the company actually bought NSWGR No. 2020, a class E10 (Z20) 2-6-4T. This was followed five years later by a second example, No. 2017. Both Beyer Peacock products dating from 1891, they were withdrawn in 1956 with the intention of overhauling No. 2017 with components salvaged from its sister, which was to be cannibalised. However, in 1955 an order was placed with Robert Stephenson & Hawthorn for a powerful 2-6-2T, with large 19 x 26 inch cylinders and 4-foot 7-inch driving wheels. Works No. E7841, it was allocated fleet No. 1 at Hebburn. After a fairly short career it was laid aside with defective

The last steam locomotive imported into Australia was Hebburn No. 1, a 2-6-2T built by Robert Stephenson, in 1955. It was withdrawn with defective cylinders just twelve years later. (Courtesy Lindsay Bridge)

cylinders in early 1967, but the intended overhaul never took place, and it was scrapped three years later. As an immediate replacement for the withdrawn No. 1, another 2-6-4T, latterly J. & A. Brown No. 26, arrived at Hebburn. No. 26, which had earlier been NSWGR No. 2013, did not prove to be a very reliable machine and was relegated to spare engine with the arrival of NSWGR 4-6-4T No. 3013 in May 1967.

In 1917 the Wickham and Bulloch Island Coal Mining Company sunk its No. 2 colliery at Cessnock, and a locomotive curiosity was employed in the construction of the rail link to the new mine. *Pygmy* was a standard six-coupled side tank built by Hudswell Clarke in 1888 for the Robinson's Beach Tramway, Sydney. When the tramway was electrified in 1900 it went to the Toronto–Fassifern tramway near Newcastle, and then passed through the hands of contractors engaged on various other construction projects. The Wickham company acquired the by now thoroughly worn-out engine in 1916. A local engineer, known as 'Duchy' Olsen, embarked on a rather quirky rebuild, and *Pygmy* entered service for its new owner powered by a petrol engine removed from a Leyland lorry! With the boiler and cylinders removed, the replacement power unit was fitted where the smokebox saddle had been located, the drive transmitted via gearbox, transfer box and cardan shaft to a new jack shaft. The side tanks, bunker and coupling rods were retained, and in its new guise the locomotive was capable of hauling two loaded wagons. In 1932 it was sent to Neath Colliery for storage.

Andrew Barclay of Kilmarnock provided the next locomotive for the Wickham company, a standard 0-4-0ST, works No. 1738 of 1923. It served for a decade before being sold to Lysaught Pty and sent to its Port Kembla works in 1933. Shortly before the departure of the Barclay, another locomotive arrived in the shape of a six-coupled saddle tank, which had been built by Vulcan Foundry in 1879. It was formerly NSWGR No. 530X but had been sold to the Commonwealth government in 1915 and subsequently employed at the naval depot at Jervis Bay. Out of use by 1938, it lay out of use for many years, partially dismantled. Although its boiler was scrapped long ago, the remains have surprisingly survived.

Although SMR passenger services had been operated by NSWGR since 1930, by the late 1950s there were serious concerns about the increasing costs of the loss-making operation. In an attempt to stem the tide, in 1958 the SMR approached the Rhodes-based rolling stock manufacturer Tulloch Ltd with regard to the feasibility of using diesel railcars. In the following year an order was placed with the firm for three such vehicles, to be delivered in 1961. The eighty-seat cars were 18.5 metres (60 feet 8 inches) long and powered by Rolls-Royce C6s FLH diesel engines with hydraulic drive, and capable of a maximum governed speed of 43 mph. Finished in a royal blue and yellow livery, the new cars were tested on the NSWGR system before delivery to East Greta. The cars, which could operate singly or as multiple units, maintained the Cessnock–Maitland passenger timetable from their introduction on 1 October 1961 until the service was withdrawn in January 1967. Then placed in store, the company made various unsuccessful attempts to sell the redundant vehicles, which were finally scrapped in 1977.

Tulloch railcar No. 1 when new in 1961. (Courtesy Lindsay Bridge)

Railcar No. 3 in 1961. (Courtesy Lindsay Bridge)

J. & A. Brown/Richmond Vale Railway Locomotives

The first two locomotives ordered by J. & A. Brown for the new line between Minmi and Hexham were 0-4-2Ts built by the Newcastle upon Tyne firm R. & W. Hawthorn. Numbered 1 and 2, they were rebuilt from side tanks to saddle tanks in 1922, in which form they continued to work for a further two decades, being withdrawn in 1942 and 1941 respectively. Kitson of Leeds supplied an 0-6-0 saddle tank in 1878, which was later joined by a similar engine, eight years its senior, which was acquired second-hand from NSWGR in 1891. Numbered 3 and 4, these two locomotives remained at work shunting Hexham staithes until as late as 1966/7, and both survive in preservation.

    A particularly interesting acquisition was made in 1905 when the company bought three 0-6-4T locomotives, which had been built by Beyer Peacock in 1885/6 for the Liverpool-based Mersey Railway, but had become redundant as a result of electrification. A fourth engine was purchased the following year, but before they left the UK the condensing apparatus, which had been fitted for working through the Mersey Tunnel, was removed, and cabs were fitted. Renumbered 5–8 by their new owner, engines Nos 5–7 retained their original names, *The Major*, *Liverpool* and *Connaught*. Nos 6–8 were all withdrawn in 1934, but although No. 5 was overhauled in 1941, it only remained at work until the following year when it too was condemned with boiler problems. However, it remained extant and is now preserved at Thirlmere, although not operational.

Above: Avonside 0-6-0ST No. 1 of 1911 rests at the back of Hexham shed, 19 January 1962. (Courtesy the late Weston Langford)

Below: Kitson 0-6-0ST No. 3 of 1878 lies out of use at the River Yard, Hexham wagon works, January 1969. (Courtesy Graeme Kaufman)

Above: J. & A. Brown No. 6 *Liverpool* was formerly Mersey Railway No. 7, seen at Pelaw Main in 1910. (Courtesy University of Newcastle)

Below: Former Mersey Railway 0-6-4T, J. & A. Brown No. 5 was out of use at Hexham in November 1967. (Courtesy the late Weston Langford)

LOCOMOTIVES CONSTRUCTED BY KITSON & CO. LIMITED.

TANK LOCOMOTIVE FOR AUSTRALIAN RAILWAY.

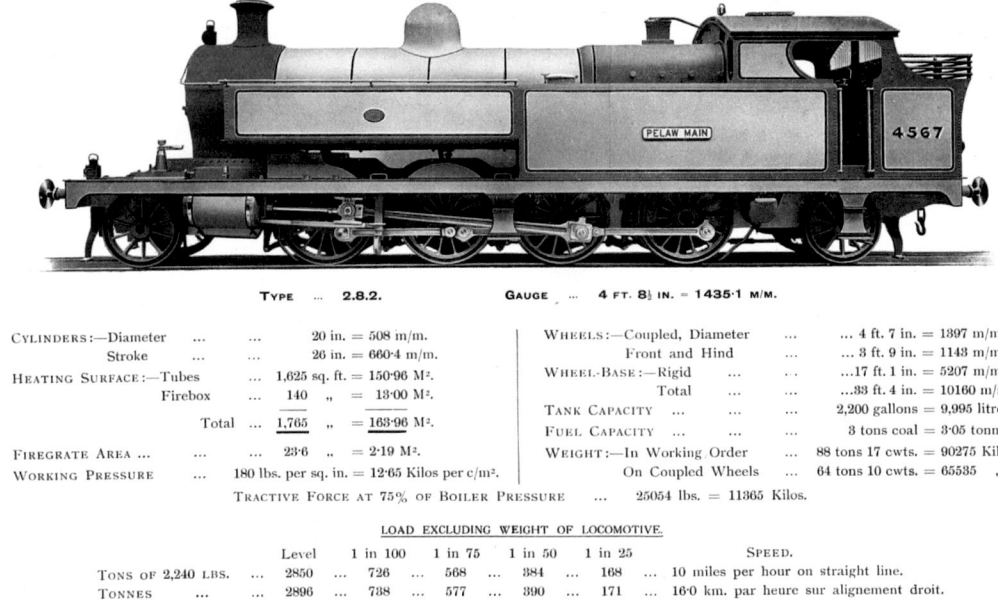

| TYPE | 2.8.2. | | GAUGE | 4 FT. 8½ IN. = 1435·1 M/M. |

CYLINDERS:—Diameter		20 in. = 508 m/m.	WHEELS:—Coupled, Diameter		... 4 ft. 7 in. = 1397 m/m.
Stroke		26 in. = 660·4 m/m.	Front and Hind		3 ft. 9 in. = 1143 m/m.
HEATING SURFACE:—Tubes		1,625 sq. ft. = 150·96 M².	WHEEL-BASE:—Rigid		...17 ft. 1 in. = 5207 m/m.
Firebox		140 „ = 13·00 M².	Total		...33 ft. 4 in. = 10160 m/m.
Total		1,765 „ = 163·96 M².	TANK CAPACITY		2,200 gallons = 9,995 litres.
			FUEL CAPACITY		3 tons coal = 3·05 tonnes.
FIREGRATE AREA ...		23·6 „ = 2·19 M².	WEIGHT:—In Working Order		88 tons 17 cwts. = 90275 Kilos.
WORKING PRESSURE		180 lbs. per sq. in. = 12·65 Kilos per c/m².	On Coupled Wheels		64 tons 10 cwts. = 65535 „
		TRACTIVE FORCE AT 75% OF BOILER PRESSURE	...	25054 lbs. = 11365 Kilos.	

LOAD EXCLUDING WEIGHT OF LOCOMOTIVE.

		Level	1 in 100	1 in 75	1 in 50	1 in 25		SPEED.
TONS OF 2,240 LBS.	...	2850	726	568	384	168	...	10 miles per hour on straight line.
TONNES	...	2896	738	577	390	171	...	16·0 km. par heure sur alignement droit.

No. 9 *Pelaw Main*, built in 1908, as featured in a contemporary Kitson catalogue. (Courtesy Alon Siton)

The company returned to Kitson for its next order, in 1908, specifying a 2-8-2 side-tank locomotive, based on a Great Central Railway eight-coupled tender class. Suitably impressed, two more were built in 1911, and the trio was numbered 9–11. Although No. 11 was withdrawn in 1949, Nos 9 and 10 were still used for heavy 'main-line' workings throughout the 1970s, with the older engine hauling its final trains in November 1980.

During the First World War the Railway Operating Division in Britain had adopted a former Great Central 2-8-0 tender locomotive class as a standard heavy freight design. Orders were placed with a number of builders, and over 500 were built in total, with many shipped to mainland Europe. Following the armistice, J. & A. Brown bought a batch of thirteen now-redundant engines, most of which had worked for a period in mainland Europe, which arrived in Australia between 1925 and 1927. Although three were transported as complete locomotives, the rest were dismantled for shipment aboard J. & A. Brown's own Glasgow-built ship, the SS *Minmi*, on its maiden voyage. The last one to be reassembled finally entered service in 1931. Some were fitted with the Westinghouse air brake, while others relied on a locomotive steam brake. There were also differences between boilers, with some superheated, but others not. Numbered 12–24, by 1970 just four remained in traffic, with the last example, No. 24, withdrawn in June 1973. Happily, three have been preserved.

The search for an additional locomotive in 1948 resulted in the purchase of a 2-6-4T, which had originally been built by Beyer Peacock in 1885 as NSWGR No. 2013 but had

Above: A study of ROD 2-8-0 No. 17 at Stockrington, 24 March 1965. (Courtesy the late Weston Langford)

Below: J. & A. Brown No. 26 became Hebburn No. 2 at Weston for a brief period in 1967. It was originally NSWGR No. 2013, new in 1885. (Courtesy the late Weston Langford)

already spent several years in industrial use. Following a heavy overhaul it was put to work, renumbered 26, and was finally withdrawn in December 1967.

Shortly before the withdrawal of No. 26, another locomotive was acquired from NSWGR in May 1967, but this time it was a 4-6-4T, No. 3013, built by Beyer Peacock in 1903. It worked at Hebburn until 1972 and then served as a spare engine for a further four years at Hexham until it too was withdrawn.

Carriages

The first carriages acquired for the new passenger services were an assortment of four-wheelers, bought from NSWGR in 1904. They were the products of various local builders, including Braid, Russell, Hudson and Mayes & Donald, with the oldest, built by Wright, dating from 1857. More four-wheelers, and several six-wheel vehicles, arrived from the state operator between 1906 and 1914, and one of the last acquisitions was a veritable antique four-wheel, first-class coupe built in 1855 and already nearly sixty years old.

In July 1907 it was reported that the East Greta company had placed an order for three 'American carriages of the most approved style' in an attempt to improve passenger facilities. Meadowbank Manufacturing Company, based near Sydney, was awarded the contract for the construction of the new bogie vehicles, which were delivered the following year, numbered in the carriage fleet as 21–23. No. 22 featured first- as well as second-class accommodation. They immediately proved so successful and a welcome upgrade from the existing stock that three more were ordered in 1909, but this time from the Australian Engineering and Rolling Stock Company, numbered 24–26. Two were brake seconds while the third, No. 25, was a composite. Two further composites, Nos 28 and 29, were delivered by the Clyde Engineering Company in 1910, along with No. 27, which provided only second-class facilities. The company was evidently pleased with Clyde Engineering as they were favoured with the next order for three more vehicles, Nos 32–34, the following year.

No further additions to the fleet were made until 1925 when, once again, Clyde Engineering was awarded a contract to supply two new second-class and one composite carriages, numbered 43, 45 and 44 respectively. A more unusual part of the order was for a bogie brake, No. 46, which was equipped with a dog box and coffin chamber!

Several early carriages ended their days converted for departmental use, for example No. 7, a four-wheel composite built by Russell in 1873, became a mess/tool van with the South Maitland breakdown train.

In 1915 the Abermain Coal Company sought permission for a passenger service, although mainly intended for the conveyance of its own employees to and from work. Two four-wheel tramcars had been obtained from the New South Wales Tramways Department (formerly Nos 193B and 196B) several years earlier, and these were joined by two more bogie tramcars from the Saywell Tramway at Rockdale, which had recently been declared redundant. The four-wheelers were numbered 1 and 2 while the ex-Saywell cars became 3 and 4, and all were adapted with suitable couplings to work with the company's locomotives.

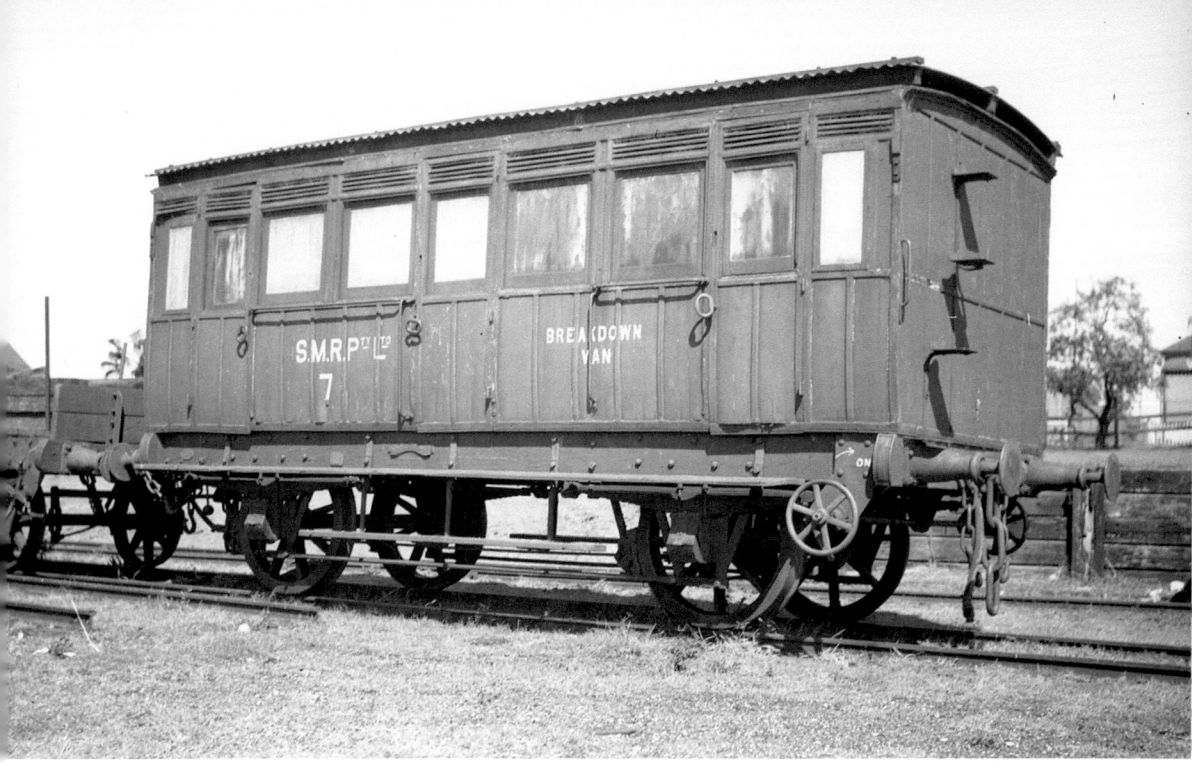

Four-wheel former composite carriage in use as van No. 7 in the SMR breakdown train. (Courtesy University of Newcastle)

Two more trams, which became Nos 5 and 6, were obtained in late 1922, and worked semi-permanently coupled, with side glazing removed and the windows partially boarded over; such basic facilities were evidently considered sufficient for commuting miners! In 1928 the old bogie tramcars were heavily rebuilt, with frames now at conventional buffing height and the appearance of box vans with open doorways positioned centrally and at each end.

Another miners' service was provided for a time between Weston and Hebburn No. 2 colliery, once again using old tramcars, but this had ended, and the vehicles disposed of, by 1939. Another oddity at Hebburn in the early 1920s was a Continental motor car, adapted for railway use! This latter, however, was not the only road vehicle adapted for railway use, as in the late 1930s a Cadillac, dating from 1923, was similarly adapted for use by JABAS management on the RVR system. Although withdrawn by 1949, it later passed to the Sydney Tramway Museum, but was unfortunately destroyed in a fire in 2017.

The railways operated for many years with large fleets of privately owned four-wheel hopper wagons, all unfitted and worked with three-link loose couplings. All of a similar basic design, with a capacity of around 10–12 tons, most were wooden-bodied with variations in wooden or steel underframes, although some were fitted with metal hoppers. In 1952 a number were fitted with air brakes to enable them to work on the NSWGR system. From 1976 government-owned air-braked bogie wagons, with capacities of up to

East Greta brake van No. 620. (Courtesy University of Newcastle)

55 tons, were introduced on to the SMR, but these were fitted with automatic buckeye-type couplings. Gradually, the 10 class 2-8-2Ts were fitted with the new couplings, and their buffers removed, to work with new stock. No. 25 was the last to be converted in 1983, as it was working at Hexham where the old-type four-wheel wagons were still in widespread use.

5

The Heritage Era

Although steam operation ended at the SMR in 1983, the coal trains continued to operate, although using diesel locomotives contracted from the state network. However, the East Greta workshops were still required to provide four steam locomotives on a daily basis for the remaining Richmond Vale operation between Stockrington and Hexham. All the non-operational locomotives were kept in store at East Greta, although one of the 2-8-2Ts, No. 19, was already 'preserved' and on display with several wagons at Port Waratah.

In 1982 No. 19 was stopped awaiting overhaul when Nos 17 and 31 were involved in the derailment at Kurri Kurri. The unfortunate No. 19 was partially cannibalised for parts, including the leading pony truck, to repair No. 31, and never ran again.

Following the closure of Richmond Main power station in 1976, the site was acquired by Cessnock City Council, and three years later the Richmond Vale Preservation Co-operative Society Ltd was formed, with the aim of preserving a part of the region's industrial railway heritage. At this time steam working could still be found elsewhere on the RVR/SMR system, but by 1980 volunteers had started relaying track at the former Richmond Colliery site, with the intention of reopening the line to Pelaw Main as the Richmond Vale Railway Museum (RVRM). The first limited diesel-hauled operation commenced in 1984. Prior to No. 19's withdrawal, the RVRM and others had made approaches to Coal and Allied to preserve one of the 10 class 2-8-2Ts, but without success.

However, as the last working locomotives in Australia, the 10 class attracted widespread interest and attention, and the National Trust of Australia gave the entire fleet protected listing. In 1989, after the cessation of regular steam at Hexham, Coal and Allied donated four engines, Nos 22, 24, 25 and 30. Arriving at Richmond early in July, No. 24 was steamed at its new home just a few weeks later. No. 25 was returned to steam following attention to its firebox and the two 2-8-2Ts became the mainstay of the heritage operation following the introduction of passenger trains between Richmond and Pelaw Main in 1991. By this time the RVRM had built a carriage shed, and Cessnock City Council had also funded the construction of a visitor entry building.

With four locomotives now safely housed at the RVRM, and No. 19 plinthed at Waratah, ownership of the remaining nine members of the class was transferred to the Hunter

Valley Training Company (HVTC), which was established in 1981 to promote employment, apprenticeship and other training opportunities in the region, and now owns the former SMR workshops at East Greta.

Three years after the end of steam at the SMR, a commemorative festival was held at Maitland in 1986, which then established itself as a prominent annual event in enthusiasts' diaries, known as the Hunter Valley Steamfest. The 1989 show even featured former LNER No. 4472 *Flying Scotsman*, which at that time was touring Australia. Over the years Steamfest has become a major attraction in the area, drawing 50,000 visitors by 2017, with a semi-permanent showground established near Maitland station. Wider interest in road steam-traction engines and other vintage vehicles led to the formation of the Maitland Steam and Antique Machinery Association, which actually manages the showground. Several former SMR 10 class locomotives have attended along with many preserved ex-main-line examples. Perhaps a unique feature and highlight of Steamfest is a race between a steam locomotive and a Tiger Moth aeroplane. In 2016 the race actually involved four trains and four aircraft! The Covid-19 pandemic caused the cancellation of the show in 2020 and 2021, and plans for the 2022 event were thwarted by flooding.

Of the nine locomotives transferred to HVTC, Nos 10 and 18 were displayed at the 1990 Steamfest, having been overhauled by Friends of the South Maitland Railways. At the following year's event both were presented in steam at East Greta along with No. 17, which had only been granted a two-day boiler certificate for the weekend. In 1994 No. 18 and ex-NSWGR C30 class 4-6-4T No. 3112 hauled a series of special trains on the SMR between East Greta and Neath, while No. 10 pottered around the yard offering brake-van rides. The group responsible for the preservation of NSWGR Pacific No. 3801, and which also operated the 'Cockatoo' heritage train between Wollangong and Moss Vale, was then offered No. 18 on permanent loan. To work these specials it was fitted with various modifications including an extended bunker, hopper ash pan and a self-cleaning smokebox. When the 'Cockatoo' trains ended in the late 1990s, No. 18 was stored at Eveleigh Works, but by 2007 it had been overhauled by an external contractor and sent to the railway workshops at Rothbury.

Rothbury had become the home of the seven remaining HVTC locomotives, Nos 17, 20, 23, 26, 27, 28 and 31, when they were sold to a Mr Chris Richards in 1990. Although No. 17 was steamed very occasionally at Ayrfield Colliery and No. 23 was later overhauled by the Friends of the South Maitland Railways, the others remained more or less untouched, although the boiler and tanks of No. 27 were used in the overhaul of No. 10. In April 2013 all seven were sold on again, to the Dorrigo Steam Railway and Museum. The Dorrigo site is essentially a large private collection which can trace its origins back to an ambitious 1973 scheme to preserve the 40-mile-long Glenreagh to Dorrigo line. Following a period of internal wrangling part of the line became the Glenreagh Mountain Railway, with the collection of locomotives and rolling stock managed by Keith Jones established at Dorrigo. Over forty steam locomotives, the earliest dating from 1878, are stored on site, many in more or less as withdrawn condition. The museum stock list records the price paid for the seven SMR engines as A$90,000 each.

Meanwhile, in 2012 there were proposals to use No. 18 for a series of special trains in the Sydney area, but mechanical issues with it and subsequently with No. 10, the back-up locomotive, led to the idea being abandoned.

At the RVRM No. 30 was returned to steam in 2000, but withdrawn in 2009, the year in which No. 19 and its wagons were donated to the museum and removed from Waratah. By now all the SMR engines in its care required expensive heavy overhauls. No. 22 was dismantled and declared beyond economic repair, with Nos 24 and 30 later dismantled for assessment. Subject to funding, it was hoped that one would be in steam by 2025.

The RVR for many years had a habit of storing withdrawn locomotives around the railway, sometimes awaiting an overhaul which never materialised. Inevitably, some were eventually scrapped, while various other examples have survived into the preservation era, having lain derelict for many years. In 1973 Coal and Allied decided that it was time for a clear-out of redundant assets and invited tenders for the disposal of assorted old locomotives.

Perhaps the most remarkable survivor from the J. & A. Brown era is No. 4, former Mersey Railway 0-6-4T No. 5, which was withdrawn with boiler problems as long ago as July 1942. It is now part of the collection at the NSW Rail Museum, Thirlmere, following rescue in 1973, although for many years subsequently it remained stored outside in unrestored condition. The oldest locomotive to be salvaged was 1878 Kitson-built J. & A. Brown No. 3, an 0-6-0ST which was stored at Rhondda Colliery for some years before removal to the Dorrigo collection in 1986. Also to be found there is another 0-6-0ST, J. & A. Brown No. 2, an Avonside dating from 1912. J. & A. Brown No. 27 is a smaller 1900-built Avonside four-coupled saddle tank, originally East Greta No. 2, also now stored at Dorrigo.

Although seven of the ROD 2-8-0s were scrapped by Sims Metal in September 1973, one, No. 21 (although according to some sources actually No. 23), was donated for preservation and has been cosmetically restored at the RVRM. Two others, Nos 20 and 24, were privately purchased and eventually found their way to Dorrigo in 1986 after periods in storage elsewhere.

The RVRM is also home to two of the Kitson-built 2-8-2Ts, Nos 9 and 10, which were moved to the fledgeling heritage line by road in June 1982. No. 9 had been the last of the trio at work, being retired in 1980.

Former East Greta No. 14, one of three 0-8-2Ts supplied by Avonside between 1908 and 1911, is also preserved at the Dorrigo site. The engine had been sold by the SMR, as successor to the East Greta company, to the Hetton Bellbird Coal Company in 1936. When withdrawn in 1972, it was donated for preservation, and like all other locomotives at Dorrigo, it is kept in outside storage, but occasionally oiled externally.

Former NSWGR No. 3013, a class S636 (later class C30) 4-6-4T built in 1903, was withdrawn in 1967 but then found itself working for a few more years in the coal industry at Hebburn. In 1982 it was sold to a private owner for preservation and subsequently dismantled at the Lachlan Valley Railway. There it remained until the parts were donated to the Australian Railway Historical Society and moved to the Canberra Railway Museum for storage. Unfortunately, the museum found itself in financial difficulties in 2016, and the remains of No. 3013 were sold on to another private owner.

Other than the RVRM heritage line between Richmond and Pelaw Main, the rest of the J. & A. Brown/RVR network has been completely dismantled, although much of the formation is still discernible with tunnels and most bridges still intact. In December 2021 planning approval was granted to convert the Hexham to Minmi trackbed into the

Richmond Vale Rail Trail, the first stage of a 25-mile-long 'multi-purpose shared pathway'. The RVRM normally offers heritage train services on three Sundays each month, and in addition to the railway interest a mining museum has been established at the Richmond Main site. The railway suffered a major setback in September 2017 when a bushfire destroyed some of its rolling stock.

A brief mention was made earlier of the Portland cement works' steam operation, and three locomotives from that company have survived. Andrew Barclay 0-6-0Ts No. 3 is part of the Dorrigo collection, while sister locomotive No. 5, which had been partly dismantled for an overhaul before withdrawal, has had a rather more chequered history since. In 1986 it was acquired by a group in the neighbouring state of Victoria, which planned to regauge it to 5 feet 3 inches for a proposed tourist line at Tallangata. After many years in storage minus wheels it was sold by auction in 2010, and was eventually moved to Dorrigo in August 2021. Perhaps more significantly, the 1892-built former NSWGR 2-6-2 saddle tank No. 2605 has been preserved at the State Mine Museum at Lithgow.

The South Maitland Railway, however, continues to exist as a commercial enterprise, and celebrated its centenary with steam locomotives Nos 10 and 18 in attendance at East Greta in 2018, but the last coal trains ran on 18 March 2020. Yancoal, the Chinese owner of the Austar mine (formerly Pelton), the last colliery still in production, had announced earlier in the year that coal output would be suspended, with a skeleton staff retained for care and maintenance of the facility, and the final train brought an end to over 120 years of coal haulage at East Greta.

The line between East Greta Junction and Cessnock remains, and as recently as 2008–09 a section of track at Loxford, near Kurri Kurri, was realigned with a new bridge built over the Hunter Expressway road scheme. The SMR, owned by Aurizon Holdings, Pty, Ltd since March 2022, offers stabling and testing facilities on its track, repair and maintenance of plant and wagons, as well as staff training.

The very end of commercial steam – No. 25 held hostage, October 1987. (Courtesy Warren Dibb)

Appendix A

Locomotive Summary

East Greta Coal Mining Company/South Maitland Railway

No.	Builder	Type	Works No.	Built	In Service	Withdrawn	Notes
1	Manning Wardle	4-4-0T	39	1862	1895	1911	Ex-NSWGR 6N
1 (2nd)	Avonside	0-8-2T	1596	1911	1911	1937	Sold Bulli Colliery
2	Avonside	0-4-0ST	1415	1900	1901	1934	Sold to J. & A. Brown, preserved Dorrigo
3	Avonside	0-6-0ST	1436	1902	1902	1928	Sold Sydney & Suburban Metal Co.
4	Kitson	0-6-0	2118	1877	1902	1922	
5	Kitson	0-6-0	2299	1879	1903	1918	Scrapped 1927
6	Avonside	0-8-0ST	1464	1903	1904	1930	
7	Kitson	0-6-0	2029	1879	1904	1922	Scrapped 1927
8	Avonside	0-6-0ST	1487	1904	1905	1928	Sold Gunnedah Colliery
9	Avonside	0-8-0ST	1481	1904	1905	1935	Sold Mount Kembla Colliery
10	Vale and Lucy	0-6-0	10	1873	1906	1911	Ex-NSWGR 21N
10	Beyer Peacock	2-8-2T	5520	1911	1912	1987	Preserved Hunter Valley Training Co.
11	Beyer Peacock	4-4-2T	1629	1877	1907	1935	Ex-NSWGR 84

No.	Builder	Type	Works No.	Built	In Service	Withdrawn	Notes
12	Beyer Peacock	4-4-2T	1628	1877	1907	1935	Ex-NSWGR 83
13	Avonside	0-8-2T	1541	1908	1908	1930	Sold 1944
14	Avonside	0-8-2T	1559	1909	1909	1930	Sold Hetton Bellbird Coal Co., preserved Dorrigo Steam Railway and Museum
15	Beyer Peacock	4-6-4T	5603	1912	1912	1965	Scrapped 1973
16	Beyer Peacock	4-6-4T	5638	1912	1912	1946	Scrapped 1973
17	Beyer Peacock	2-8-2T	5790	1914	1914	12/1983	Preserved Dorrigo
18	Beyer Peacock	2-8-2T	5909	1915	1915	12/1984	Preserved Hunter Valley Training Co.
19	Beyer Peacock	2-8-2T	5910	1915	1915	11/1981	Preserved Richmond Vale Railway
20	Beyer Peacock	2-8-2T	5998	1920	1920	02/1985	Preserved Dorrigo
21	Beyer Peacock	4-4-2T	3335	1891	1919	1935	Ex-NSWGR M51, sold to NSWPWD 1941
22	Beyer Peacock	2-8-2T	6055	1921	1921	09/1987	Preserved Richmond Vale Railway
23	Beyer Peacock	2-8-2T	6056	1921	1921	07/1980	Preserved
24	Beyer Peacock	2-8-2T	6125	1922	1922	09/1987	Preserved Richmond Vale Railway
25	Beyer Peacock	2-8-2T	6126	1923	1923	09/1987	Preserved Richmond Vale Railway
26	Beyer Peacock	2-8-2T	6127	1923	1923	07/1983	Preserved Dorrigo
27	Beyer Peacock	2-8-2T	6137	1923	1923	03/1987	Preserved Dorrigo
28	Beyer Peacock	2-8-2T	6138	1923	1923	12/1983	Preserved Dorrigo

No.	Builder	Type	Works No.	Built	In Service	Withdrawn	Notes
29	Beyer Peacock	4-6-4T	6139	1923	1923	1965	Scrapped 10/1973
30	Beyer Peacock	2-8-2T	6294	1926	1926	09/1987	Preserved Richmond Vale Railway
31	Beyer Peacock	2-8-2T	6295	1926	1926	06/1984	Preserved Dorrigo

J. & A. Brown

No.	Builder	Type	Works No.	Built	In Service	Withdrawn	Notes
1	R. & W. Hawthorn	0-4-2T	947	1856	1859	1942	Rebuilt as saddle tank 1922
2	R. & W. Hawthorn	0-4-2T	948	1856	1859	1941	Rebuilt as saddle tank 1922
3	Kitson	0-6-0ST	2236	1878	1878	1966	Preserved Dorrigo
4	Kitson	0-6-0ST	1620	1870	1891	1967	Ex-NSWGR No. 20 preserved Newcastle Museum
5	Beyer Peacock	0-6-4T	2601	1885	1905	1942	Ex-Mersey Railway No. 1 *The Major*, preserved Thirlmere
6	Beyer Peacock	0-6-4T	2607	1886	1905	1934	Ex-Mersey Railway No. 7 *Liverpool*, sold to Cessnock Collieries 1934
7	Beyer Peacock	0-6-4T	2782	1886	1905	1934	Ex-Mersey Railway No. 9 *Connaught*
8	Beyer Peacock	0-6-4T	2604	1885	1908	1934	Ex-Mersey Railway No. 4 *Gladstone*
9	Kitson	2-8-2T	4567	1908	1908	1980	Named *Pelaw Main*, preserved RVRM
10	Kitson	2-8-2T	4798	1911	1911	1976	Named *Richmond Main*, preserved RVRM

No.	Builder	Type	Works No.	Built	In Service	Withdrawn	Notes
11	Kitson	2-8-2T	4834	1911	1911	1949	Named *Hexham*, scrapped 1966
12	North British	2-8-0	22213	1919	1926	Scrapped 1968	ROD No. 2123
13	North British	2-8-0	22209	1919	1926	Scrapped 1973	ROD No. 2119
14	North British	2-8-0	22161	1919	1926	Scrapped 1966	ROD No. 2070
15	North British	2-8-0	21866	1918	1927	Scrapped 1973	ROD No. 1889
16	North British	2-8-0	21867	1918	1927	Scrapped 1973	ROD No. 1890
17	North British	2-8-0	21886	1918	1927	Scrapped 1973	ROD No. 1909
18	North British	2-8-0	22038	1918	1927	Scrapped 1968	ROD No. 1889
19	North British	2-8-0	21918	1918	1927	Scrapped 1973	ROD No. 1941
20	North British	2-8-0	22042	1918	1927	1968	ROD No. 1984, preserved Dorrigo
21	Kitson	2-8-0	5201	1918	1927	Scrapped 1973	ROD No. 1615, ran as No. 23 in last years
22	Great Central	2-8-0	ROD 2002	1919	1927	Scrapped 1973	
23	Great Central	2-8-0	ROD 2004	1919	1927	1971	Preserved RVRM
24	Great Central	2-8-0	ROD 2003	1919	1927	1973	Preserved Dorrigo
26	Beyer Peacock	2-6-4T	2567	1885	1948	1967	Ex-NSWGR No. 2013, scrapped 1970
3013	Beyer Peacock	4-6-4T	4456	1903	1967	1976	Ex-NSWGR No. 3013, preserved

Hebburn Locomotives

No.	Builder	Type	Works No.	Built	In Service	Withdrawn	Notes
62XX	Robert Stephenson	0-6-0	2195	1874	1903	1927	NSWGR, scrapped Hebburn
1	Robert Stephenson	2-6-2T	E7841	1955	1955	1967	Scrapped 1970
2	Beyer Peacock	0-6-0ST	2575	1884	1903	1938	Transferred from Australian Agricultural Co., Newcastle, scrapped 1947
3	Beyer Peacock	0-6-0ST	4558	1903	1903	1920	Transferred from Australian Agricultural Co. 1918, sold Lithgow Iron Works 1920
26	Beyer Peacock	2-6-4T	2567	1885	1967	1967	Ex-NSWGR 2013, scrapped 1970
2017	Beyer Peacock	2-6-4T	3289	1891	1939	1956	Ex-NSWGR 2017
2020	Beyer Peacock	2-6-4T	3206	1891	1934	1955	Ex-NSWGR 2020
3013	Beyer Peacock	4-6-4T	4456	1903	1967	1976	Ex-NSWGR 3013, hired from RVR, tr. to Hexham 1972, preserved by NSWPR South Maitland

Abermain Coal Company

No.	Builder	Type	Works No.	Built	In Service	Withdrawn	Notes
1	Avonside Engine Co.	0-6-0ST	1606	1911	1911	1961	To RVR, Hexham 1955, scrapped 1966
2	Avonside Engine Co.	0-6-0ST	1916	1922	1922	1969	To RVR, Hexham 1963, preserved Dorrigo

No.	Builder	Type	Works No.	Built	In Service	Withdrawn	Notes
1	Hudswell Clarke	0-6-0T	290	1886	1920	1925	Named *Saywell*, sold Howley & Co., Glenrock Railway

Cessnock, Wickham & Bullock Island Collieries

No.	Builder	Type	Works No.	Built	In Service	Withdrawn	Notes
	Pygmy Hudswell Clarke	0-6-0T	296	1888	1916	1961	Converted with Leyland petrol engine
——	Andrew Barclay	0-6-0ST	1738	1922	1923	1933	Sold Lysaght, Port Kembla, preserved Richmond Vale Railway
530X	Vulcan Foundry	0-6-0ST	834	1879	1932	1961	Ex-NSWGR 530X, remains preserved Yass Tramway
	Liverpool Beyer Peacock	0-6-4T	2607	1866	1934	1943	Ex-J. & A. Brown No. 6

Appendix B

Leading Dimensions of Principal Locomotive Classes

South Maitland Railway

Class	10	NSWGR C30 (formerly S636 design)
Fleet nos	10, 17–20, 22–28, 30, 31	15/16, 29
Builder	Beyer Peacock	Beyer Peacock
Works nos	5520, 5790, 5909/10/98, 6055/6, 6125-7, 6137/8, 6294/5	5603/38, 6139
Date built	1911–25	1912, 1923
Wheel arrangement	2-8-2T	4-6-4T
Cylinders	20 x 26 inches	18.5 x 24 inches
Coupled wheels	4 feet 3 inches	4 feet 7 inches
Leading wheels	2 feet 9 inches	3 feet 1 inches
Trailing wheels	2 feet 9 inches	3 feet 1 inches
Wheelbase	31 feet 4 inches overall	34 feet 8 inches
Grate area	23.91 square feet	24 square feet
Total heating surface	1,838 square feet	1,450 square feet
Boiler pressure	180 lb	160 lb
Water capacity	2,400 gallons	1,580 gallons
Coal capacity	4 tons	3.25 tons
Weight in working order	83 tons 10 cwt	73 tons

J. & A. Brown/Richmond Vale Railway

Class		ROD
Fleet nos	9–11	12–24
Builder	Kitson	North British (9) Kitson (1) Great Central Railway (3)
Works nos	4567/4798/4834	NBL: 21866/7/86/21918/22038/42/22161/209/213 Kitson: 5201
Date built	1908–11	1917–18
Wheel arrangement	2-8-2T	2-8-0
Cylinders	20 x 26 inches	21 x 26 inches
Coupled wheels	4 feet 7 inches	4 feet 8 inches
Leading wheels	3 feet 9 inches	3 feet 6 inches
Trailing wheels	3 feet 9 inches	4 feet 4 inches (tender)
Wheelbase	33 feet 4 inches overall	25 feet 5 inches (engine only)
Grate area	23.60 square feet	26.25 square feet
Total heating surface	1,625 square feet	1,756 square feet
Boiler pressure	180 lb	180 lb
Water capacity	2,200 gallons	4,000 gallons
Coal capacity	3 tons	6 tons
Weight in working order	88 tons 17 cwt	74 tons 13 cwt (engine only)

Bibliography

Andrews, Brian, *Coal Railways and Mines: The Story of the Railways & Collieries of J & A Brown*, Iron Horse Press, 2004/7

Andrews, Brian, *Coal, Railways & Mines: The Railways & Collieries of the Greta & South Maitland Coalfields, Volumes 1–5*, Brian Andrews, 2017 et seq.

Attenborough, Peter, *South Maitland Railways,* Eveleigh Press, 2001

Cooke, David et al., *Coaching Stock of the NSW railways Volume 2*, Eveleigh Press, 2002

Delany, J. W., 'The South Maitland Railways', *Australian Railway Historical Society Bulletin*, July 1966, pp. 145–64

Driver, R., 'Steam Operations on the SMR', *Australian Railway History*, January 2004, pp. 28–39

Eardley, Gifford H., *The Railways of the South Maitland Coalfields,* Australian Railway Historical Society, 1969

McNicol, Steve, *Coals to Maitland* Railmac Publications, 1982

Preston, Ron, *The Richmond Vale Railway*, NSW Rail Transport Museum, 1989

Acknowledgements

I would like to thank all those who have so generously allowed me to use their photographs in this book. In particular, I would like to acknowledge Dennis Rittson, Lindsay Bridge, Ian Lynas, Warren Dibb, Brian Ayling, Graeme Skeet, Graeme Kaufman, Frank Hinde and the late Weston Langford. Maitland Library and the University of Newcastle have also kindly permitted the use of images from their collections.

I would also like to thank Ed Slee and Jeff Mullier for answering my queries, and I am further grateful to Jeff for the map of the railway systems.